THE NEW VIKING CHILDREN'S WORLD ATLAS

An Introductory Atlas for Young People

Jacqueline Tivers and Michael Day

VIKING

VIKING
Published by the Penguin Group
Penguin Books USA Inc., 375 Hudson Street, New York, New York 10014, U.S.A.
Penguin Books Ltd, 27 Wrights Lane, London W8 5TZ, England
Penguin Books Australia Ltd, Ringwood, Victoria, Australia
Penguin Books Canada Ltd, 10 Alcorn Avenue, Toronto, Ontario, Canada M4V 3B2
Penguin Books (N.Z.) Ltd, 182-190 Wairau Road, Auckland 10, New Zealand

Penguin Books Ltd, Registered Offices: Harmondsworth, Middlesex, England

This revised edition published by Viking, a division of Penguin Books USA Inc., 1994

10 9 8 7 6 5 4 3 2 1

Library of Congress Cataloging-in-Publication Data

Tivers, Jacqueline.
The new Viking children's world atlas: an introductory atlas for
young people / Jacqueline Tivers and Michael Day.
    p. cm.
Rev. ed. of : The Viking children's world atlas. 1985
Summary: An introductory atlas focusing on the political and
natural geography, as well as the industries and resources, of
each of the different regions of the world.
ISBN 0-670-85481-6
    1.Children's atlases   [1.Atlases]   I. Day, Michael, 1938-
    ill. II. Tivers, Jacqueline. Viking children's world atlas.
    III. Viking Press   IV. Title
G1021.T5826 1994 <G&M>
912-dc20 CIP MAP AC   93-41434

Printed in Hong Kong
Set in Helvetica

Acknowledgments
The publishers wish to thank Robert Harding Associates, Colour
Library International, Spectrum Colour Library, British Petroleum,
C. Peter Kimber, Pica Design, Lawrence R. Lowry/Louis Mercier, J. Allan
Cash and David Muench for providing photographs for use in this
edition.

Contents

Page

4-5	Introducing Sam and Sarah
6-7	World Map
8-9	Shading and Symbols
10-13	United States of America
14-15	Canada
16-17	Mexico and Central America
18-19	South America
20-21	Western Europe
22-23	Eastern Europe
24-25	Scandinavia
26-27	Russian Federation
28-29	Africa
30-31	Middle East
32-33	Southwest Asia
34-35	China and Its Neighbors
36-37	Southeast Asia
38-39	Japan and New Zealand
40-41	Australia
42-43	North and South Polar Regions
44-45	The World from Space
46	The Ages of Life on Earth
47	Fast Facts

Introducing Sam and Sarah

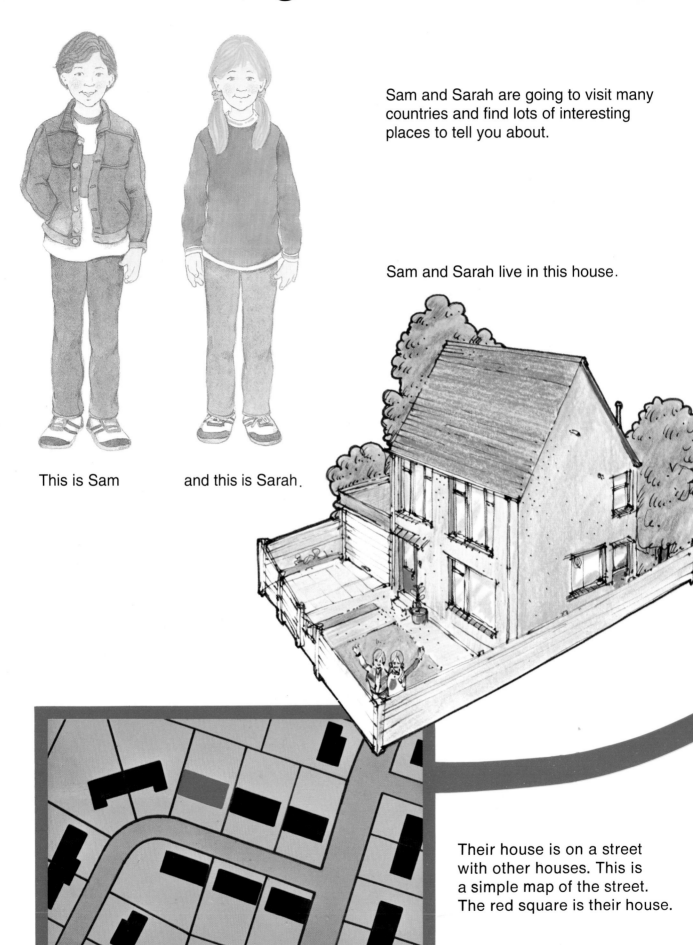

Sam and Sarah are going to visit many countries and find lots of interesting places to tell you about.

Sam and Sarah live in this house.

This is Sam   and this is Sarah.

Their house is on a street with other houses. This is a simple map of the street. The red square is their house.

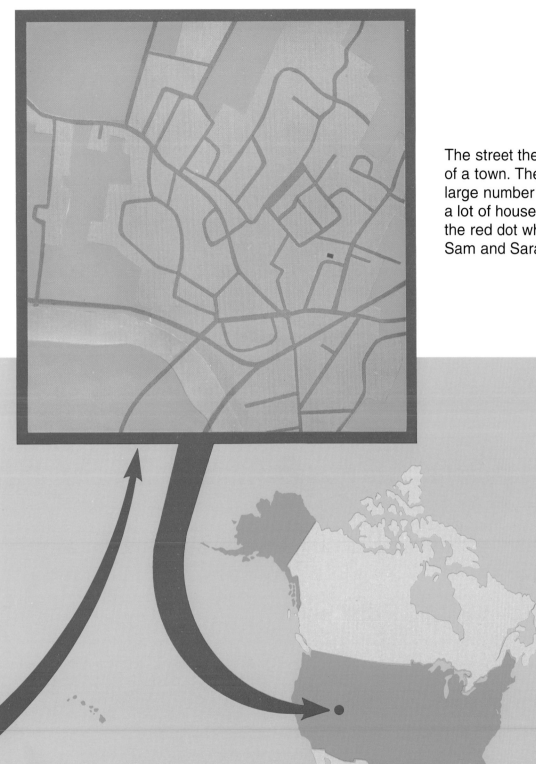

The street they live on is part of a town. The town has a large number of streets and a lot of houses. Can you find the red dot which shows where Sam and Sarah live?

Their town is in the United States, which is a large country. The red dot shows where the town is on this map. As you can see on the map, the United States goes from one ocean to the other. It also has two other parts. The state of Alaska is in the far North. The state of Hawaii is far out in the Pacific Ocean. Can you find the United States on the map on pages 6 and 7? On pages 10, 11, 12, and 13 there are bigger maps of the United States.

World Map

On this map of the world, we can see the country where Sam and Sarah live. The United States is shaded in red.

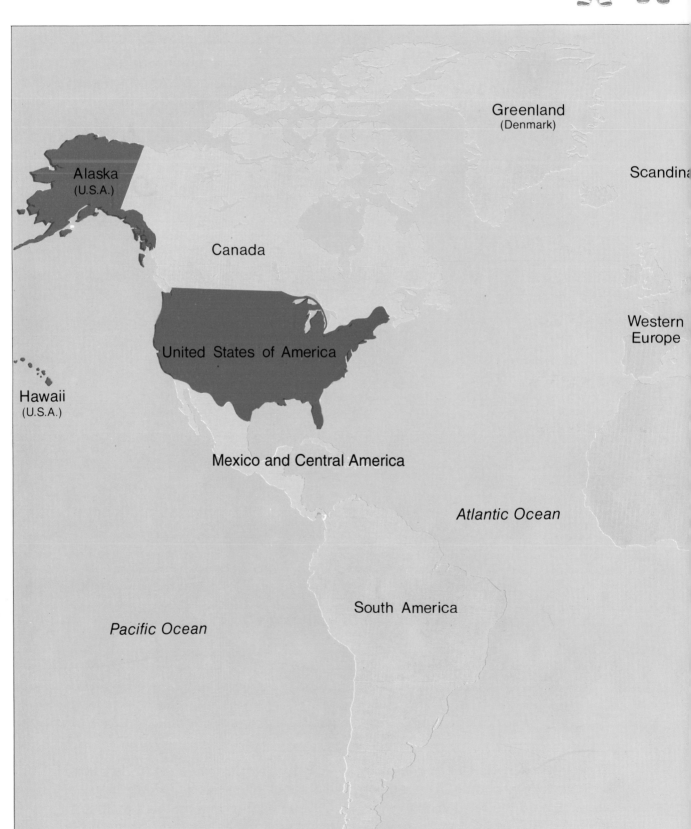

Greenland
(Denmark)

Scandina

Alaska
(U.S.A.)

Canada

Western
Europe

United States of America

Hawaii
(U.S.A.)

Mexico and Central America

Atlantic Ocean

South America

Pacific Ocean

Each part of the world that Sam and Sarah are going to visit is shown, too.

On every page of the book you will find a small map of the world. It shows in red the area that Sam and Sarah are visiting.

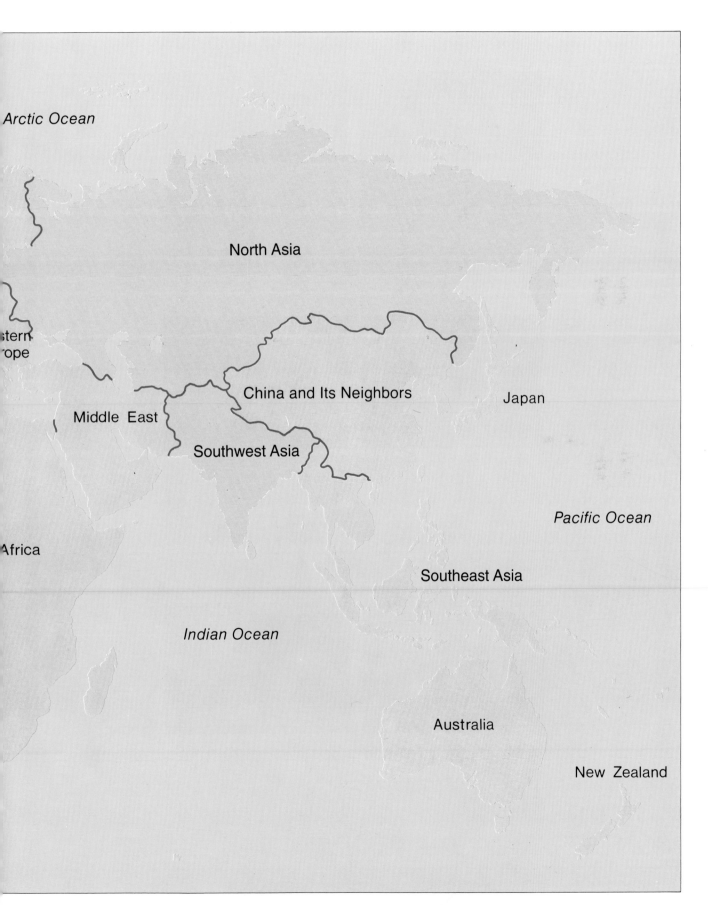

Arctic Ocean

North Asia

stern
ope

China and Its Neighbors

Japan

Middle East

Southwest Asia

Pacific Ocean

Africa

Southeast Asia

Indian Ocean

Australia

New Zealand

Of course, the continents of the world are not really blank as they appear on the last map. Some parts of the world are covered with thick forests. Other parts are desert or farmland. In some areas there are high mountains and deep valleys. As Sam and Sarah travel around the world they will see all these different types of land, and the maps of the places they visit will be shaded in different colors.

Grassland

Forest

Farmland

Mountains

Desert

Tundra

Sam and Sarah will also see some of the things people do in the countries they visit. The symbols on the maps will show where people work, the crops they grow, and the animals they keep.

These are the symbols you will find on the maps.

The crops people grow

 timber

 wheat and barley

 corn

 rice

 potatoes and yams

 peanuts or groundnuts

 apples and pears

 oranges and lemons

 grapes

 bananas

 dates

 sugar

 coffee

 tea

 cocoa

 cotton

 rubber

 tobacco

 palm oil

The animals people keep

 cows (Europe and America)

 cows (Africa and Asia)

 sheep (Europe and Australia)

 sheep (Africa and Asia)

 pigs

Where people work

 offices

 factories

 mining

 coal mining

 oil and gas

 nuclear power

 sea port

 fishing port

Places for holidays

 beach vacations

 skiing vacations

 hiking vacations

Look for Sam and Sarah on every map and see what they are doing in each part of the world.

On most pages you will also find some wild animal symbols, such as lions, birds, and elephants. These show where those animals live.

9

United States of America

Native American tribes once wandered freely all over what is now called the United States. Then other people came to live in America from many different parts of the world. Today the United States has modern farms, small towns, big cities, and many different kinds of industry.

Many cities have tall buildings, called sky-scrapers, where people work in offices. The buildings are tall because land in the cities costs a lot of money. It is cheaper to build upwards than outwards. This map shows the continental United States. On pages 12 and 13 there is a map of all the states.

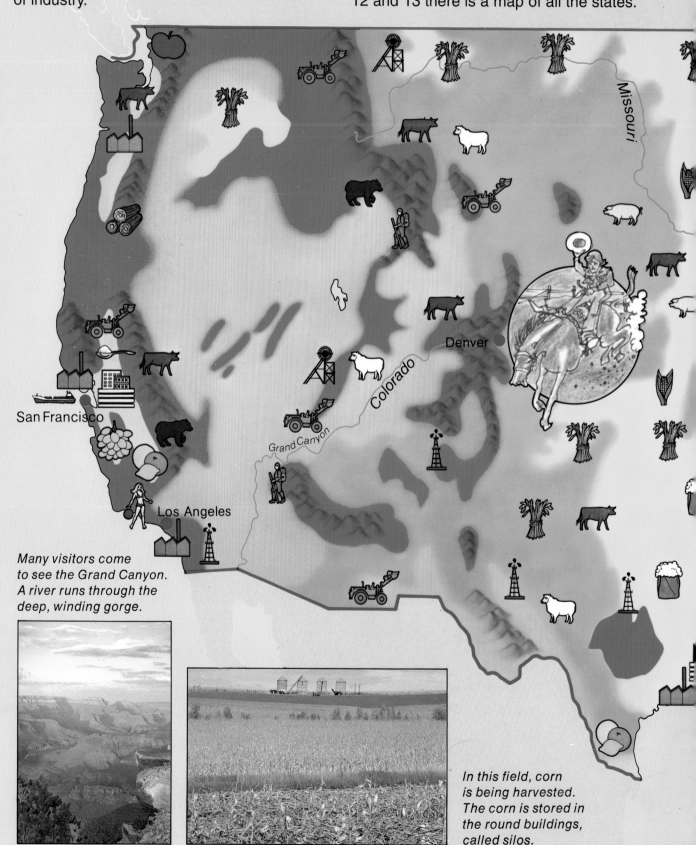

Missouri

Denver

Colorado

Grand Canyon

San Francisco

Los Angeles

Many visitors come to see the Grand Canyon. A river runs through the deep, winding gorge.

In this field, corn is being harvested. The corn is stored in the round buildings, called silos.

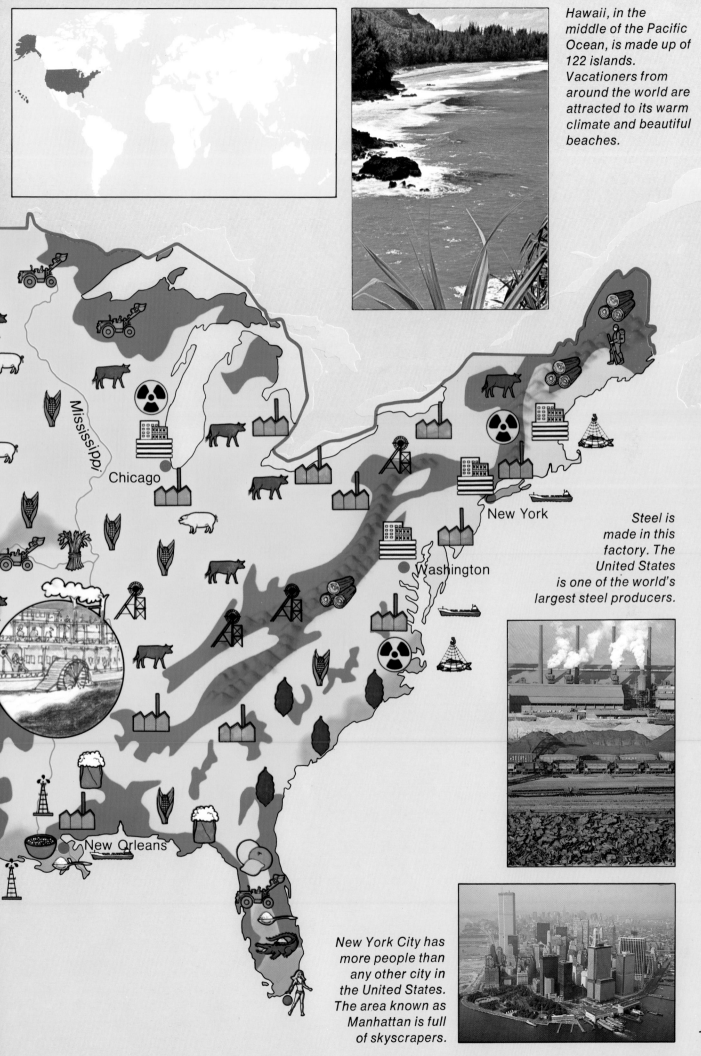

Hawaii, in the middle of the Pacific Ocean, is made up of 122 islands. Vacationers from around the world are attracted to its warm climate and beautiful beaches.

Mississippi

Chicago

New York

Washington

New Orleans

Steel is made in this factory. The United States is one of the world's largest steel producers.

New York City has more people than any other city in the United States. The area known as Manhattan is full of skyscrapers.

11

United States of America

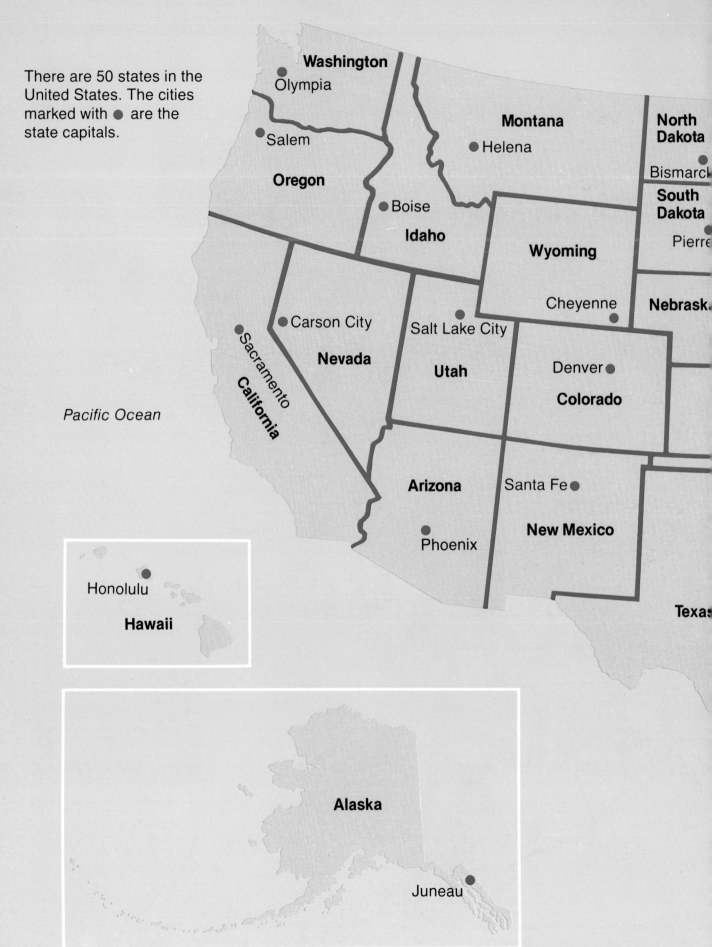

There are 50 states in the United States. The cities marked with ● are the state capitals.

Washington
Olympia

●Salem

Oregon

Montana
●Helena

●Boise

Idaho

Wyoming

North Dakota
●Bismarck

South Dakota
Pierre

Cheyenne●

Nebraska

●Carson City

Sacramento
California

Nevada

Salt Lake City●

Utah

Denver●

Colorado

Pacific Ocean

Arizona

Santa Fe●

New Mexico

●Phoenix

Honolulu●

Hawaii

Texas

Alaska

Juneau●

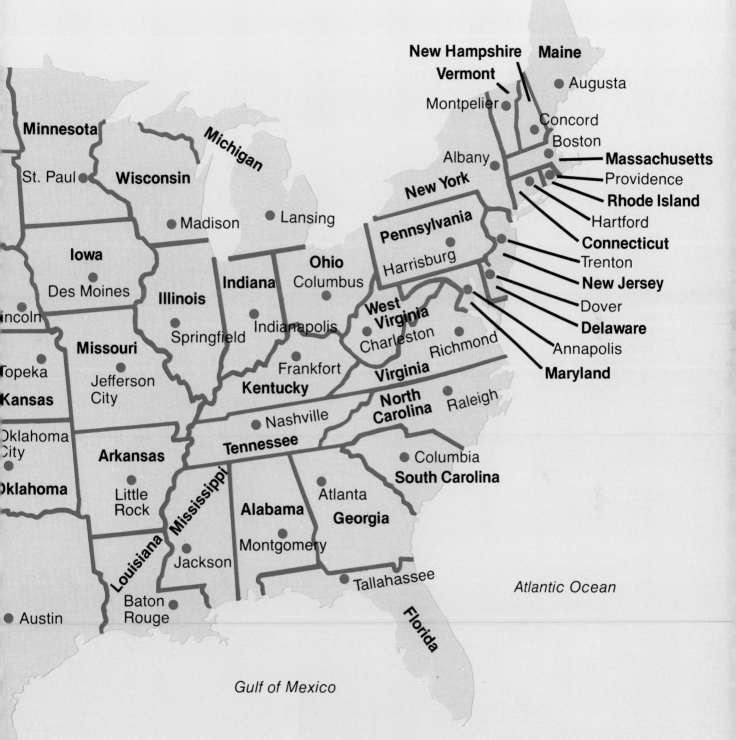

Minnesota

St. Paul

Michigan

Wisconsin

Madison

Lansing

Iowa

Des Moines

Illinois

Ohio

Columbus

Indiana

Pennsylvania

Harrisburg

New York

Albany

New Hampshire

Vermont

Montpelier

Maine

Augusta

Concord

Boston

Massachusetts

Providence

Rhode Island

Hartford

Connecticut

Trenton

New Jersey

Dover

Delaware

Annapolis

Maryland

Lincoln

Springfield

Indianapolis

West
Virginia

Charleston

Richmond

Topeka

Missouri

Jefferson
City

Frankfort

Virginia

Kentucky

North
Carolina

Raleigh

Kansas

Nashville

Tennessee

Oklahoma
City

Arkansas

Columbia

South Carolina

Oklahoma

Little
Rock

Mississippi

Alabama

Atlanta

Georgia

Louisiana

Jackson

Montgomery

Tallahassee

Atlantic Ocean

Austin

Baton
Rouge

Florida

Gulf of Mexico

Canada

There is tundra in northern Canada, in an area where it is too cold for trees to grow. In winter the land is covered with snow and ice. In the summer most of the ice melts and small plants grow everywhere in the wet ground. Not many people live in the tundra or the great forests. The cities and farms are further south where it is warmer.

Alaska
(U.S.A.)

Mackenzie

Vancouver

The mountain scenery in Canada is very beautiful. These are the Rocky Mountains.

The man sawing the tree is called a lumberjack. The trees will be made into paper. Strong machines are used to lift the trees onto huge trucks and boats.

Niagara Falls is a very big waterfall near the Great Lakes. Electric power is produced by using the water to turn machinery.

Hudson Bay

anada

Quebec
Montreal
Ottawa
Toronto

Wheat is grown in huge fields in the Prairies. Big machines are used to cut the wheat and take out the grain for food.

Toronto is the capital of Ontario, Canada. This is a picture of the City Hall.

15

Mexico and Central America

In this area, there are high, dry mountains and wet, tropical jungles. Most people work on farms. Sugar cane and bananas are grown nearly everywhere. In the Caribbean Sea there are many tropical islands with palm trees along the shore.

Brightly colored fish live around the coral reefs in the sea. Can you find the Panama Canal on the map? It is used by big ships to cross from one ocean to another.

Gulf of Mexico

Mexico

Mexico City

Guatemala

El Salvador

Many centuries ago, the Mayans built impressive temples to show their importance and power

Many people come to this busy market to shop.

Bananas grow in big bunches. Most of them are sold to other countries. When they are shipped they are green. They ripen to a bright yellow color while they travel.

16

Palm trees grow on this tropical island.

This is the cane from which sugar is obtained.

Some of the islands are famous for their steel bands. The people play drums made from oil drums, which give a very lively sound.

The Bahamas

Nassau

Havana

Cuba

Jamaica Kingston

Haiti

Dominican Republic

Puerto Rico

Caribbean Sea

onduras

Nicaragua

Barbados

Trinidad

Costa Rica

Panama City

Panama Canal

South America

The inland part of Brazil is covered by a huge tropical forest. It rains nearly every day and the trees grow very tall and close together. Hardly any sunlight reaches the ground. Millions of plants and animals live in the rain forests. In South America there are some big farms whose owners are very wealthy. But most people are poor and have to work hard just to feed their families.

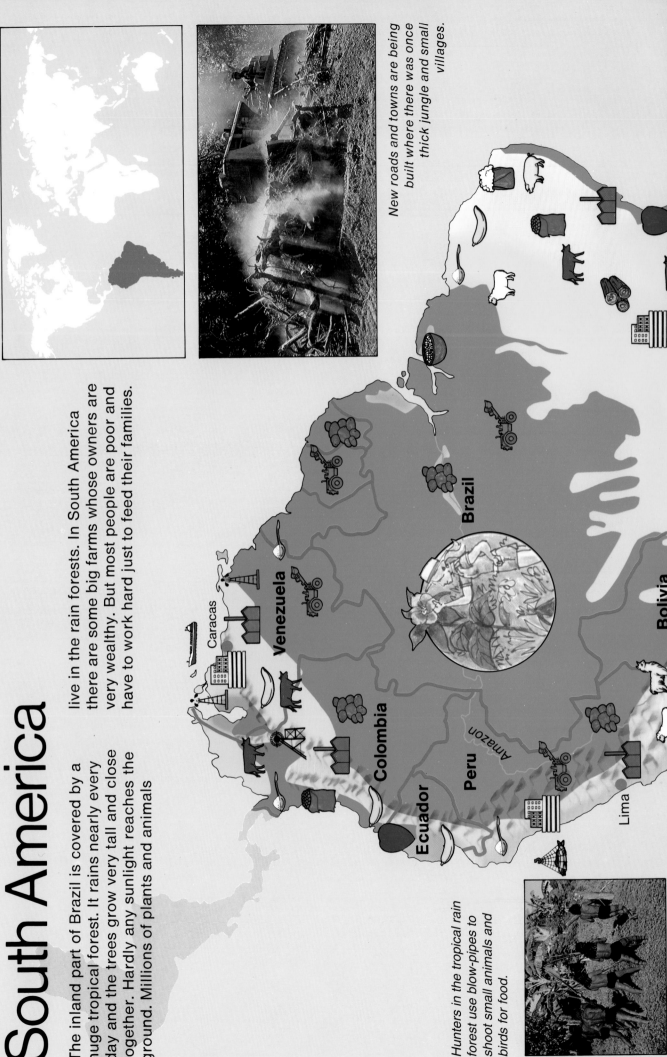

New roads and towns are being built where there was once thick jungle and small villages.

Hunters in the tropical rain forest use blow-pipes to shoot small animals and birds for food.

Caracas

Venezuela

Colombia

Ecuador

Peru

Amazon

Brazil

Bolivia

Lima

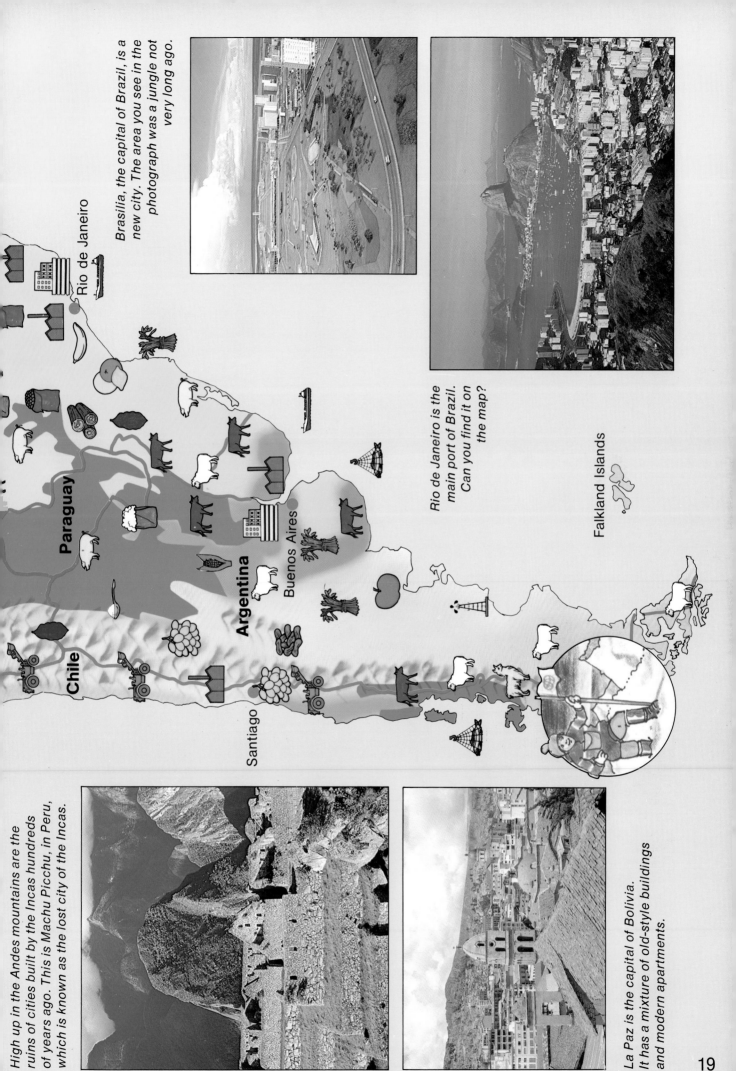

Brasilia, the capital of Brazil, is a new city. The area you see in the photograph was a jungle not very long ago.

Rio de Janeiro

Rio de Janeiro is the main port of Brazil. Can you find it on the map?

Paraguay

Argentina

Buenos Aires

Chile

Santiago

Falkland Islands

High up in the Andes mountains are the ruins of cities built by the Incas hundreds of years ago. This is Machu Picchu, in Peru, which is known as the lost city of the Incas.

La Paz is the capital of Bolivia. It has a mixture of old-style buildings and modern apartments.

19

Western Europe

There are many areas with lots of factories.
Near the coast big steel works and oil
refineries have been built, but in the
mountains the factories are smaller.
In Switzerland watches and clocks are
made. The Swiss have become famous
as manufacturers of good watches.
Because the south of Europe is warm
and sunny, grapes and other fruit can
be grown. The coast of southern
Europe is a popular vacation area.

*Rotterdam is one of the biggest
ports in the world.*

*People in Spain wear
costumes like these for
dancing and festivals.*

The Brandenburg Gate is one of Berlin's historic landmarks

Grapes are grown in vineyards in many parts of Western Europe. They are used to make different types of wine.

North Sea

Netherlands

Hamburg

Rotterdam

Berlin

Belgium

Brussels

Germany

Rhine

This is a picture of the Matterhorn, a high mountain in the Alps.

Vienna

Austria

Switzerland

Geneva

Italy

Corsica

Rome

Sardinia

Sicily

Two thousand years ago the Romans had a very big empire. They built many cities. Many of the ruins can still be seen today. This one is the Colosseum in Rome.

This oil rig is supplying Britain with oil from under the North Sea.

Mediterranean Sea

21

Eastern Europe

The river shown on the map is the Danube. Boats travel along this river carrying goods to countries in Eastern and Western Europe.

In the cities of Eastern Europe there are many large factories and apartment complexes. In the countryside farmers grow corn, potatoes, wheat, and other crops.

Shops and offices edge the picturesque Wenceslas Square in Prague.

This is a picture of Castle Cesky Krumlov in the Czech Republic.

Christmas tree decorations are being made in this

Poland

Warsaw

Czech Republic

Prague

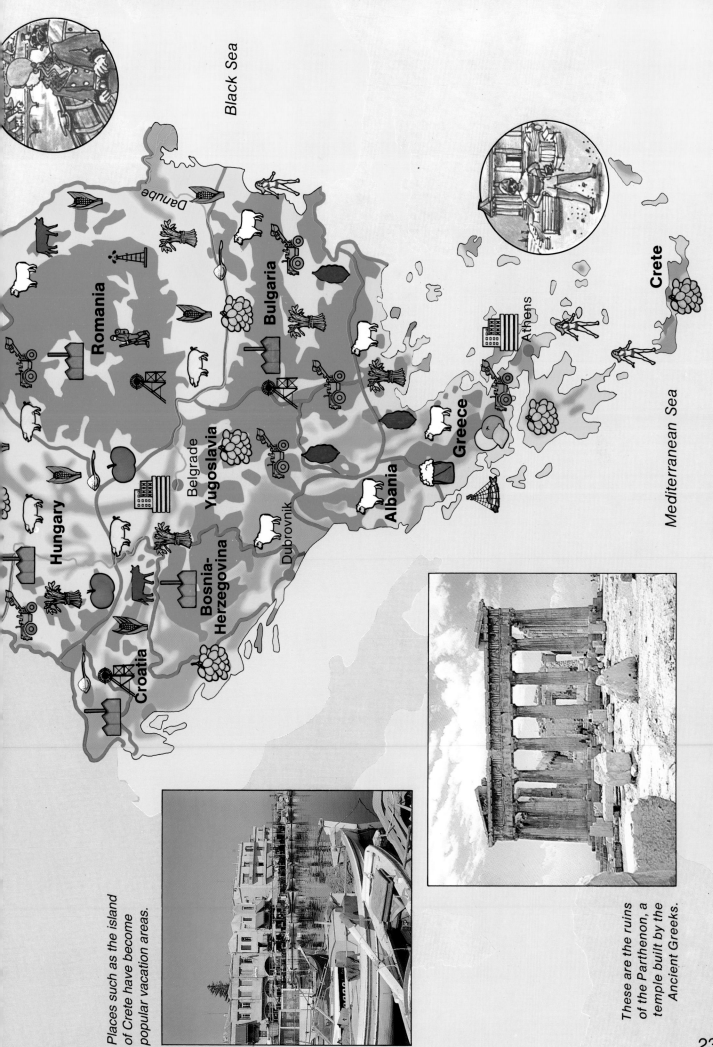

Black Sea

Danube

Romania

Bulgaria

Hungary

Belgrade
Yugoslavia

Bosnia-
Herzegovina

Croatia

Dubrovnik

Albania

Greece

Athens

Crete

Mediterranean Sea

Places such as the island of Crete have become popular vacation areas.

These are the ruins of the Parthenon, a temple built by the Ancient Greeks.

Scandinavia

Most of the land is covered with forests, lakes, and mountains. In the North it is called "The Land of the Midnight Sun." This is because in summer the sun shines all through the night. In winter it is dark all through the day. The farmers in Denmark sell butter and bacon to other countries. The people in Sweden make paper from their trees for other countries. Most people live in the South, where there are farms, factories, and towns.

The Lapps look after herds of reindeer in the far North.

There are 66,000 lakes in Finland.

This is a picture of northern Scandinavia, "The Land of the Midnight Sun."

This steep-sided valley is called a fjord. There are fjords all along the coast of Norway.

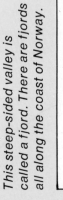

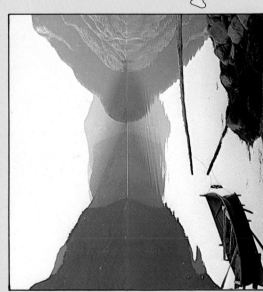

This is part of the city of Stockholm. The view is from the roof of the City Hall.

The Little Mermaid watches over the port of Copenhagen. It is a famous landmark.

Finland

Helsinki

Sweden

Stockholm

Norway

Oslo

Gothenburg

Copenhagen

Denmark

Russian Federation

The Russian Federation is the biggest country in the world. In 1991 the provinces of the Soviet Union separated into several independent countries. Most people live in the western part working in factories or on large farms. People are being encouraged to move to new towns in Siberia, the eastern part of Russia. The great forest is known as the Taiga. It is shaded green on the map. The forest stretches across the whole country.

Russian Federation

St.Petersburg

Moscow

Volga

Ukraine

Kazakhstan

Black Sea

Caspian Sea

This monastery is in Georgia, a country near the Black Sea.

At the Kosmos Center in Moscow we can see how the people have used science to change and develop their lives.

Large areas of flat grassland are known as steppes.

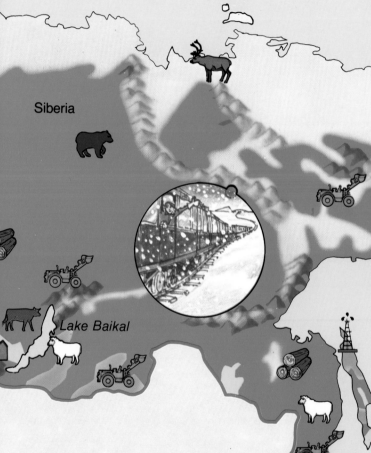

Siberia

Lake Baikal

Vladivostok

This village is by Lake Baikal, in Siberia.

This is the Moskvich car factory in Moscow.

From the River Moskva we can see the center of Moscow, the capital city of the Russian Federation. We can see its government building, the Kremlin.

Africa

There are many different countries in Africa. Some are big and others are very small. Some countries are rich and others are poor. Most of them used to be ruled by people from other countries, such as Britain and France. Now they are ruled by their own people. The Sahara in North Africa is the largest dry area of land in the world. It is over 3,000 miles from one end to the other.

In the desert there are some places where water can be found, often in small pools. Around these pools, plants and palm trees can grow.

Oil tankers and other big ships can go through the Suez Canal. This can make their journey shorter.

These elephants live in a special park in Uganda where they are protected from hunters. Can you find some wild animals on the map?

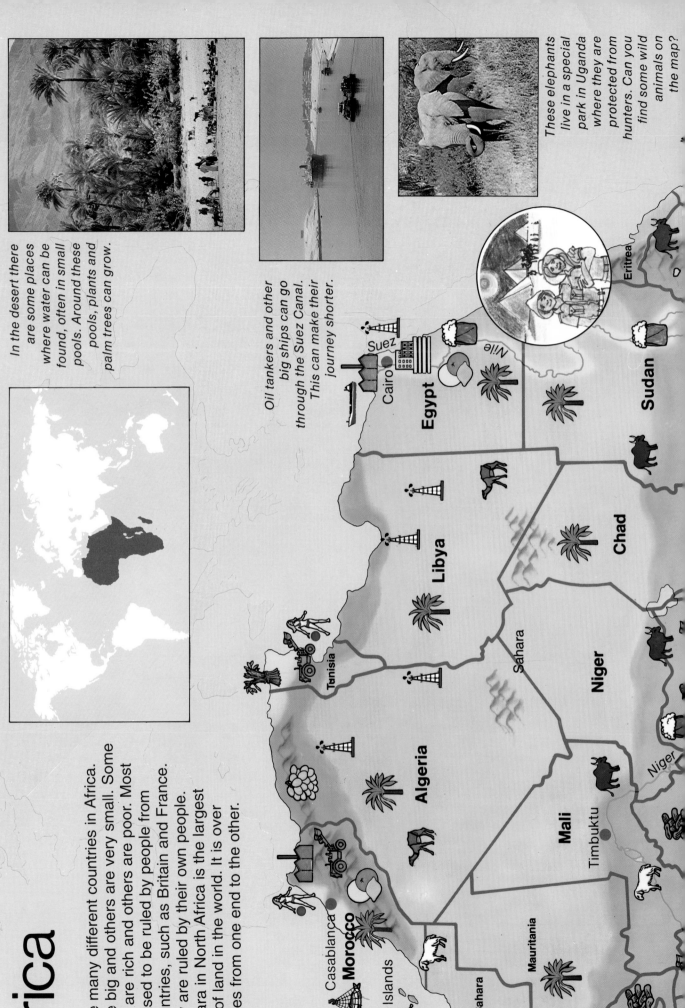

Suez
Cairo
Egypt
Nile
Sudan
Eritrea
Libya
Chad
Algeria
Niger
Tunisia
Sahara
Mali
Timbuktu
Niger
Casablanca
Morocco
Canary Islands
Western Sahara
Mauritania

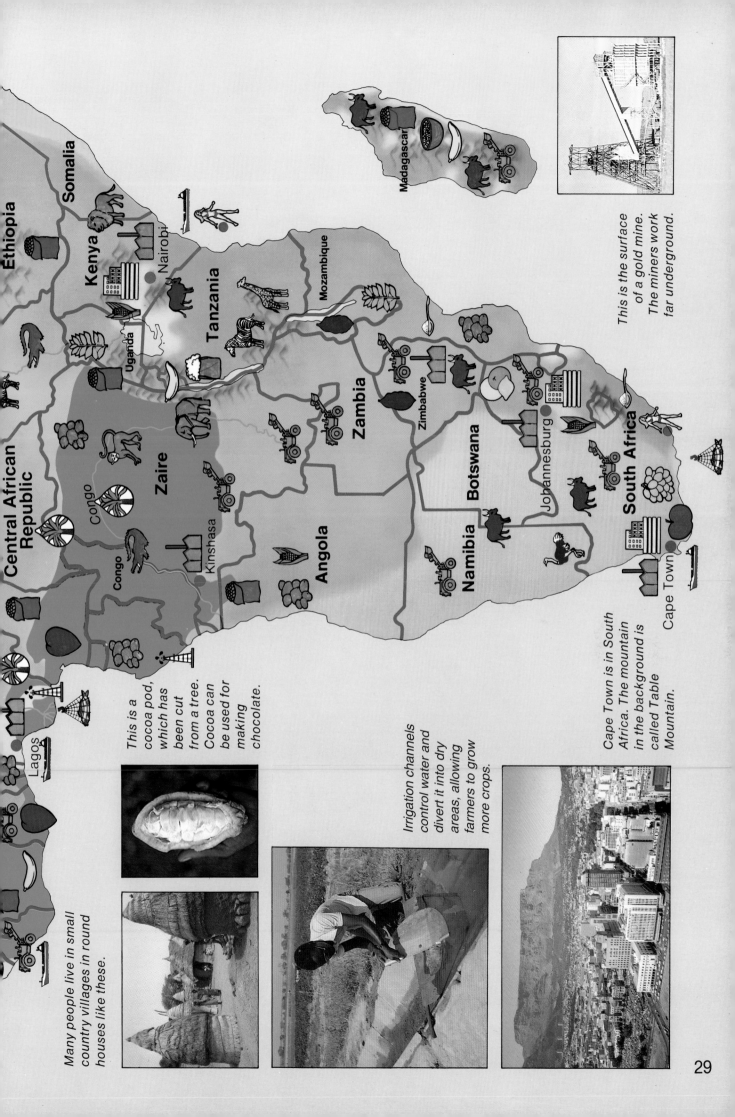

Ethiopia

Somalia

Kenya

Nairobi

Uganda

Tanzania

Mozambique

Madagascar

Central African Republic

Congo

Congo

Zaire

Kinshasa

Zambia

Zimbabwe

Johannesburg

Botswana

Namibia

Angola

South Africa

Cape Town

Lagos

This is the surface of a gold mine. The miners work far underground.

Many people live in small country villages in round houses like these.

This is a cocoa pod, which has been cut from a tree. Cocoa can be used for making chocolate.

Irrigation channels control water and divert it into dry areas, allowing farmers to grow more crops.

Cape Town is in South Africa. The mountain in the background is called Table Mountain.

Middle East

A lot of the world's oil comes from the Middle East. Most of the land is very dry. In daytime it is hot, but at night it becomes cold. Some of the people who live in this area are nomads who travel in the desert with their camels. The Holy Cities of Jerusalem and Mecca are in the Middle East. Can you find them on the map?

Istanbul

Turkey

Cyprus

Lebanon

Israel
Jerusalem

Jord

Syr

Mediterranean Sea

Red Se

Sheepskin coats and rugs are for sale on this street in Jerusalem.

Onami tribesmen protect their heads from the fierce sun by wearing turbans.

These nomads live in sturdy tents that can protect them from the heat of the day and the cold of the night.

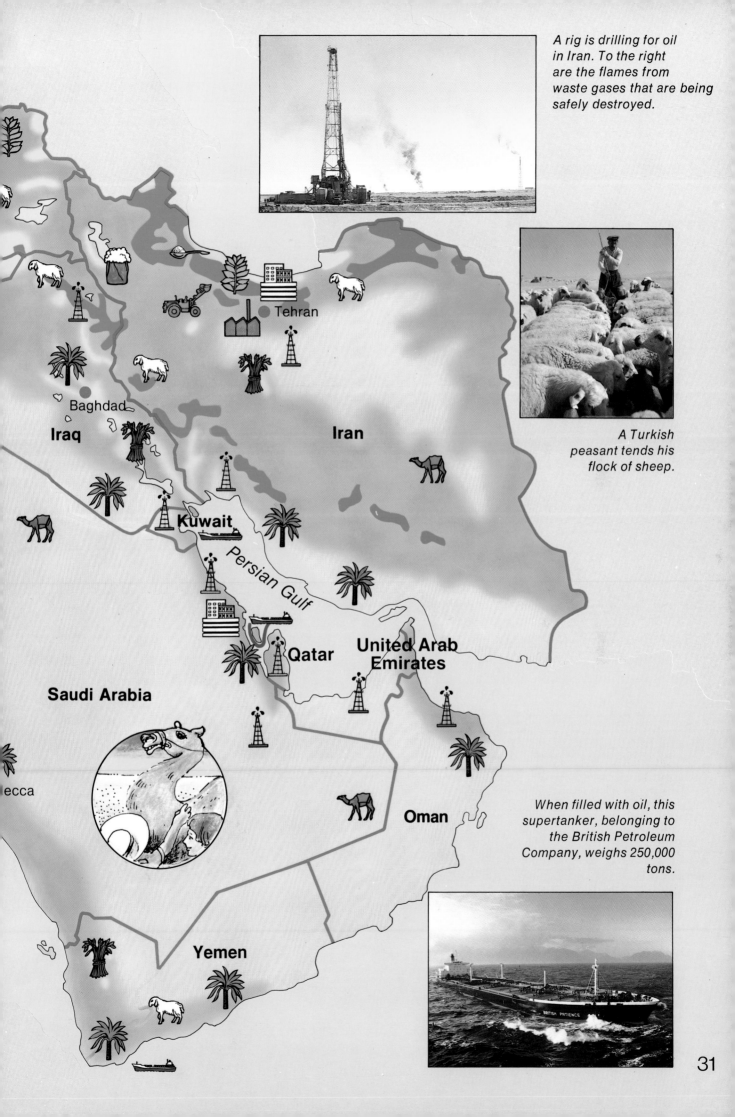

A rig is drilling for oil in Iran. To the right are the flames from waste gases that are being safely destroyed.

A Turkish peasant tends his flock of sheep.

Tehran

Baghdad

Iraq

Iran

Kuwait

Persian Gulf

Saudi Arabia

Qatar

United Arab Emirates

Mecca

Oman

When filled with oil, this supertanker, belonging to the British Petroleum Company, weighs 250,000 tons.

Yemen

Southwest Asia

This is an area of high mountains and wide valleys. Can you find Mount Everest on the map? It is the highest place on earth.

For long periods there is no rain at all and the crops die. Sometimes there are terrible floods which destroy houses and farms. Many people do not have enough to eat. In the cities there are some modern factories and new buildings, as well as beautiful temples and old palaces. India and Pakistan both have large textile industries.

The Taj Mahal was built by a great prince. It is made of white marble.

Mount Everest is the highest mountain in the world.

Mount Everest

Nepal

Delhi

Pakistan

Indus

Afghanistan

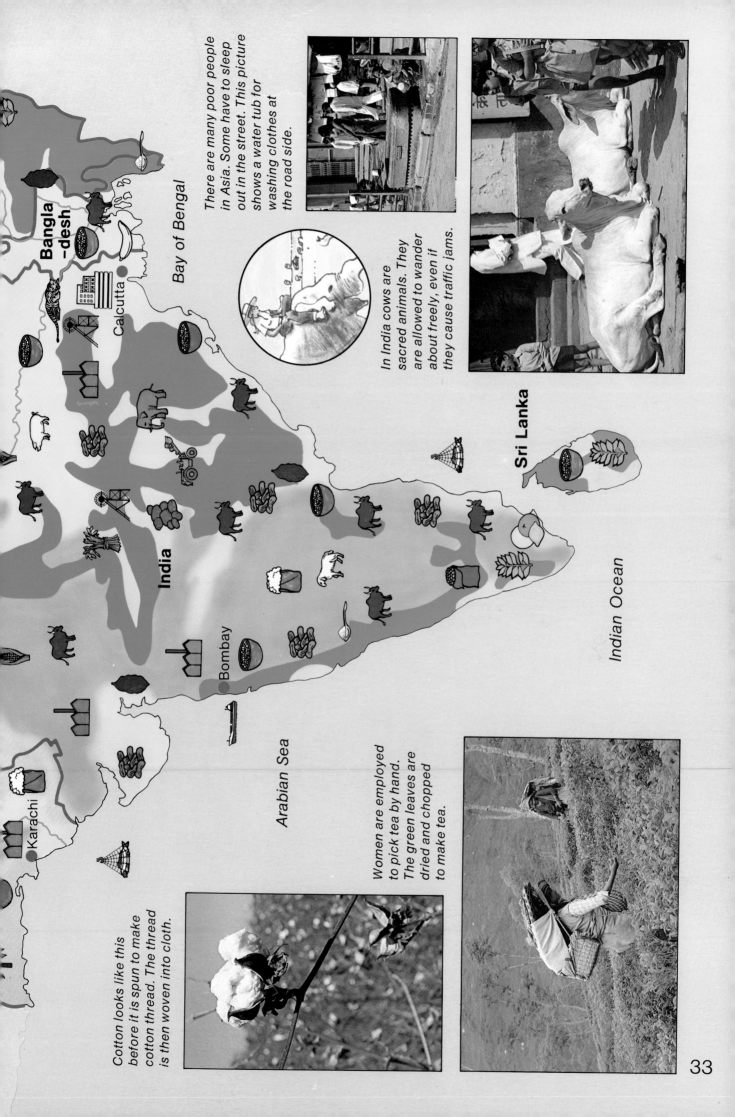

Bangla-desh

India

Calcutta

Bay of Bengal

Karachi

Bombay

Arabian Sea

Sri Lanka

Indian Ocean

There are many poor people in Asia. Some have to sleep out in the street. This picture shows a water tub for washing clothes at the road side.

In India cows are sacred animals. They are allowed to wander about freely, even if they cause traffic jams.

Women are employed to pick tea by hand. The green leaves are dried and chopped to make tea.

Cotton looks like this before it is spun to make cotton thread. The thread is then woven into cloth.

33

China and Its Neighbors

More people live in the People's Republic of China than in any other country in the world. Most of them live in the countryside and work together in the fields or in small factories. Rice is their main food. It needs a lot of water to grow, so the rivers are used to irrigate the fields.

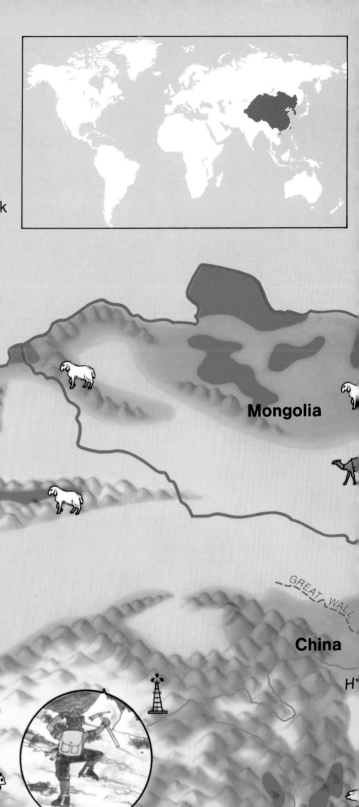

Mongolia

GREAT WALL

China

H

These Chinese women are washing clothes.

These fields are flooded to grow rice. They are called paddy fields.

34

Many people work together to dig up rocks and to build dams that control the rivers.

In this school in China, the children are writing the Chinese alphabet.

North Korea

South Korea

Peking

Yangtze Kiang

Shanghai

Taiwan

Hong Kong

The Great Wall was built a very long time ago to protect the Chinese from their enemies.

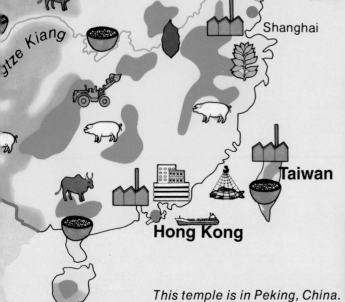

This temple is in Peking, China.

35

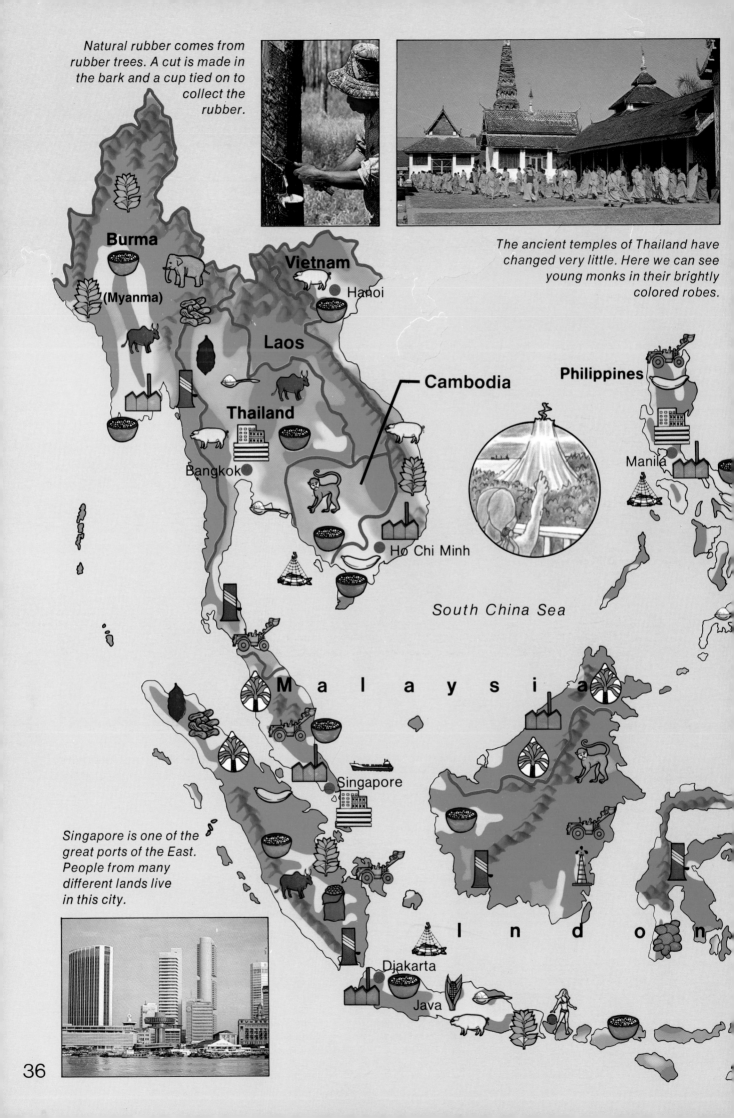

Natural rubber comes from rubber trees. A cut is made in the bark and a cup tied on to collect the rubber.

The ancient temples of Thailand have changed very little. Here we can see young monks in their brightly colored robes.

Burma

(Myanma)

Vietnam

Hanoi

Laos

Cambodia

Philippines

Thailand

Bangkok

Manila

Ho Chi Minh

South China Sea

M a l a y s i a

Singapore

Singapore is one of the great ports of the East. People from many different lands live in this city.

I n d o n

Djakarta

Java

Southeast Asia

Most of Southeast Asia is covered with tropical rain forest. Hundreds of years ago, traders came from Europe and took back home precious oils and spices. They called these islands the East Indies. Today, there are many very big cities in Southeast Asia. Millions of people leave the countryside and move to the cities to find jobs. Can you find the island of Java on the map? It is one of the most densely populated islands in the world.

The islands are crowded with people. Some of the people live in boats on the water. They buy food at floating markets like this one in the photograph.

Where the land is hilly, rice is grown on fields which are cut into the hillsides.

In the big cities, many of the poor people live in shanties. These are houses made of old scraps of wood and other things.

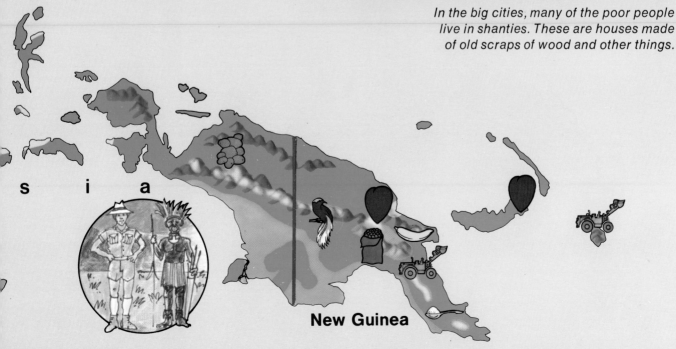

s   i   a

New Guinea

Japan

The Japanese factories make radios, calculators, automobiles, motor bikes, and computer parts. Japan also has a very large number of shipyards and builds more ships than any other country.

At this Japanese port, cuttle fish have been hung up to dry.

Tokyo

Osaka

This colorful food store is in Tokyo. As you can see from the shop sign, Japanese writing is very different from our own.

Mount Fuji is a volcano. It is quite near Tokyo and people often go to visit it.

New Zealand

New Zealand sends wool, meat, and butter to other countries. In the North Island there are volcanoes and geysers. A geyser is a hot water fountain.

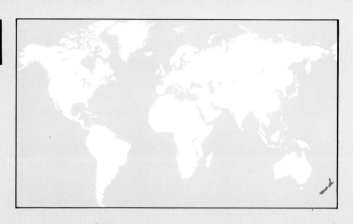

These people are Maoris. Maoris lived in New Zealand long before European explorers discovered their country.
They used to wear clothes like this all the time, but now they only dress up for special occasions.

North Island

Auckland

Wellington

Mount Cook is the highest mountain in New Zealand. There are many mountains and glaciers in the South Island.

Christchurch

South Island

This is a geyser. The water is very hot, as you can see from the steam coming off it.

39

Australia

The first people in Australia were the Aborigines. Now they have land in only a few parts of Australia. Australia is a rich country. There are large farms and mines. Nearly all the people live in towns and cities on the coast. The middle of Australia is very dry and not many people live there.

Only a few Aborigines still appear like this. Most of them wear modern clothes and live in towns.

Perth

Wool from sheep is an important product of Australian farms.

Big new mines are being dug in Australia, far away from any cities. This one will provide ore for making iron.

The Great Barrier Reef is made of coral. Coral is formed from the shell skeletons of millions of small sea animals.

A white wallaby carries its baby in its pouch.

Sydney is the biggest city in Australia. Very large ships can come into the harbor. In this photograph you can see the Harbour Bridge and, to the left of it, the Opera House.

GREAT BARRIER REEF

lice Springs

Darling

Murray

Sydney

Canberra

Melbourne

North Polar Region

If Sam and Sarah were to get on a spaceship and orbit the Earth, they would see that the Earth is round. At the extreme northern end of the Earth is the North Pole, where the sea is frozen to form a solid ice cap.

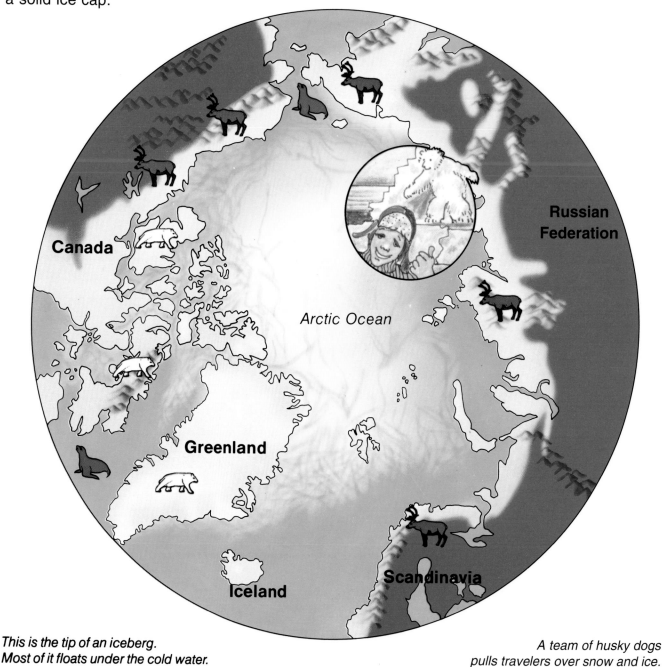

Russian Federation

Canada

Arctic Ocean

Greenland

Iceland

Scandinavia

This is the tip of an iceberg.
Most of it floats under the cold water.

A team of husky dogs
pulls travelers over snow and ice.

South Polar Region

At the other end of the Earth, Sam and Sarah would look down from their spaceship and see the South Pole and the frozen continent of Antarctica.

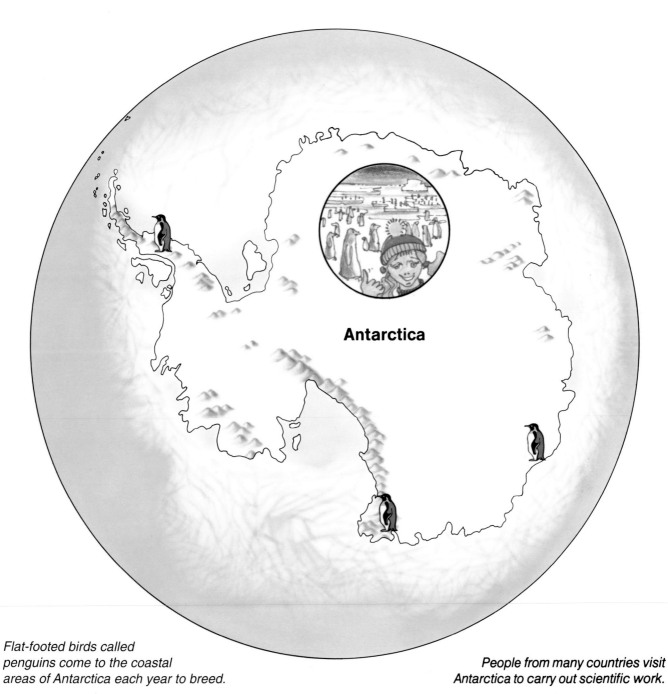

Antarctica

Flat-footed birds called penguins come to the coastal areas of Antarctica each year to breed.

People from many countries visit Antarctica to carry out scientific work.

43

The World from Space

Looking at the world from space, Sam and Sarah would see the continents partly covered by clouds, as in this photograph taken by a satellite going around the Earth. Can you see which part of the world is shown on the photograph? The area of the Earth which is turned away from the sun is in darkness. The world spins round in space every day and there is always one half of our planet where it is night, and one half where it is day.

The planets and the Earth move around the sun.

MERCURY
VENUS
EARTH
MARS
JUPITER
SATURN
URANUS
NEPTUNE
MOON
PLUTO

As the Earth travels through space around the Sun, so the Moon travels around the Earth.

The Ages of Life on Earth

The fourth age of life on earth

Some scientists have found signs of people dating back over 1 million years.

The third age of life on earth

From about 65 million years ago, the modern animals began to appear.

Giant Sloth

Mammoth

Saber-toothed Tiger

Hipparion

The second age of life on earth

Ended about 65 million years ago.

Rhamphorhynchus

Stegosaurus

Tyrannosaurus

The first age of life on earth

About 500 million years ago.

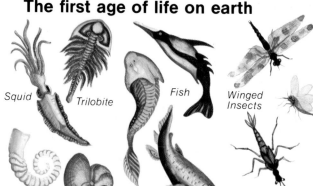

Squid

Trilobite

Fish

Winged Insects

Shellfish

Fast Facts

Facts About the World

Longest river: Nile, Africa   4,160 miles
Largest ocean: Pacific Ocean   almost 64 million square miles
Largest freshwater lake: Lake Superior, North America   32,150 square miles
Highest mountain: Mount Everest, Southwest Asia   29,028 feet
Largest country: Russian Federation   over 6.5 million square miles
Country with the most people: China   over 1 billion people
City with the most people: Mexico City, Mexico   over 20 million people

The Continents

Asia

Africa

North America

South America

Antarctica

Europe

Australia

The Oceans

Pacific

Atlantic

Indian

Arctic

Antarctic

Facts About the United States of America

Capital: Washington, D.C.
Largest state: Alaska   586,412 square miles
Smallest state: Rhode Island   1,214 square miles
Longest river: Mississippi River   3,740 miles
Population: 249 million people
Size: Over 3.5 million square miles